537

KU-072-509

SCIENCE TOPICS

Electricity & Magnetism

CHRIS OXLADE

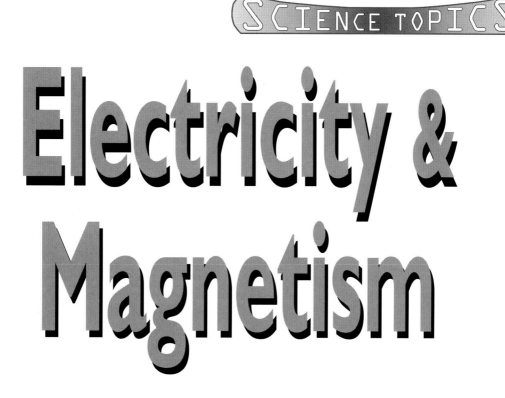

Heinemann
LIBRARY

First published in Great Britain by Heinemann Library,
Halley Court, Jordan Hill, Oxford OX2 8EJ,
a division of Reed Educational and Professional Publishing Ltd.

Heinemann is a registered trademark of Reed Educational & Professional
Publishing Limited.

OXFORD MELBOURNE AUCKLAND
JOHANNESBURG BLANTYRE GABORONE
IBADAN PORTSMOUTH NH (USA) CHICAGO

© Reed Educational and Professional Publishing Ltd 1999
The moral right of the proprietor has been asserted.

All rights reserved. No part of this publication may be reproduced, stored in a
retrieval system, or transmitted in any form or by any means, electronic,
mechanical, photocopying, recording, or otherwise without either the prior
written permission of the Publishers or a licence permitting restricted copying in
the United Kingdom issued by the Copyright Licensing Agency Ltd,
90 Tottenham Court Road, London W1P 0LP.

Designed by AMR
Illustrations by Art Construction
Printed in Hong Kong

03 02 01 00 99
10 9 8 7 6 5 4 3 2 1

ISBN 0 431 07669 3

British Library Cataloguing in Publication Data
Oxlade, Chris
Electricity and magnetism. – (Science topics)
1.Electricity – Juvenile literature 2.Magnetism – Juvenile
literature
I.Title
537

KNOWSLEY HEY
SCHOOL LIBRARY

Acknowledgements
The Publishers would like to thank the following for permission to reproduce photographs:
J. Allan Cash pg 15; Bruce Coleman pg 24; Mary Evans Picture Library pgs 5, 13; Gould, Peter,
pgs 12, 26; Trevor Hill Photos pg 25; Image Select/Ann Ronan/Photri pg 17; Science Photo
Library pgs 8 /John Walsh, 10, 19 /Peter Menzel, 27 /Publiphoto Diffusion/R. Maisonneuve;
The Stock Market pg 4; Tony Stone Images pg 28 /Mitch Kezar.

Cover photograph reproduced with permission of Image Bank/Pete Turner.

Our thanks to Jane Taylor for her comments in the preparation of this book.

Every effort has been made to contact copyright holders of any material reproduced in this book.
Any omissions will be rectified in subsequent printings if notice is given to the Publisher.

For more information about Heinemann Library books, or to order, please phone 01865 888066, or
send a fax to 01865 314091. You can visit our web site at www.heinemann.co.uk

Any words appearing in the text in bold, **like this**, are explained in the Glossary.

Contents

Electricity and magnetism

If you look around your home or school, you will find dozens of machines and devices that work using electricity. Some are quite simple, such as electric lights and food-mixers, and use electricity to create light and movement. Others are amazingly complex, such as televisions and computers – these use electricity for power and control.

Magnetism is closely related to electricity. Magnets attract materials that contain **magnetic** materials, such as iron. As well as the magnets we use to stick messages to refrigerator doors, there are many other examples in our homes, such as the magnets and **electromagnets** (magnets made by electricity) found in many electric machines.

SCIENCE ESSENTIALS

Electricity is a form of energy. Electricity and magnetism are closely related.
A wire carrying electricity attracts or repels a magnet.

Trains powered by electricity ready to travel from London to Paris.

Electricity as energy

Electricity is a form of energy. It is very convenient for us to use in our homes, offices and factories because, unlike other forms of energy, it can be changed easily into light, heat and movement. It is also a clean form of energy and is easy to send along wires to where it is needed.

A brief history of electricity

The Ancient Greeks already knew about electricity in about 600 BCE. They knew that when a substance called **amber** is rubbed with a cloth, it attracts light objects such as feathers and dust. In fact, the word electricity comes from *elektron*, the Greek word for amber. However, the Greeks did not understand the cause of this effect.

In the 1700s scientists began to work with electricity. The English physicist Stephen Gray (1666–1736) showed that electricity could flow through some substances but not others. The American statesman and scientist Benjamin Franklin (1706–90) showed that lightning was the same as **static electricity** through the risky experiment of flying a kite into thunder clouds. Electricity flowed down the string and jumped into his hand, giving him a small electric shock. He also put forward the theory of positive and negative charges, and **attraction** and **repulsion**.

More progress was made by Italian scientist Alessandro Volta (1745–1827), who realized that chemical reactions could make electricity, and devised the first **battery**, called a voltaic pile. With a supply of electricity to use, scientists went on to build electric circuits, light-bulbs and electric motors.

The place most people first saw the use of electricity was in street lighting, which began to appear in the late 1800s, soon after the first electricity generating stations were built. It was not long before domestic electricity became available, and people began to light their homes with electric lights, cook with electricity and buy electrically-powered household gadgets.

An engraving made in 1890 showing one of the first electricity generating stations, making power for New York

Static electricity

Electricity comes in two different forms. The first is **static electricity** – the type of electricity that makes your hair stand on end when you comb it, and that makes lightning during a storm. The other form of electricity is **current electricity**, which you can find out about on page 8.

you can find out about on page 8.

SCIENCE ESSENTIALS

Static electricity is created by rubbing certain materials against each other.
There are two types of electric charge – positive and negative.
Opposite charges attract each other.
Like charges repel each other.

Rubbing for static

Static electricity is normally caused by certain materials rubbing against each other. For example, when you comb your hair with a plastic comb, the comb rubs against your hair, creating static electricity.

When you take some clothes off, they can rub against each other, also creating static electricity. But why does static electricity make you hair stand up and your clothes crackle?

Electric charges

When certain materials rub against each other, tiny particles called **electrons** get transferred from the surface of one material to the surface of the other. The direction in which the electrons move depends on the two materials. For example, electrons transfer from cotton cloth to plastic, and from glass to a silk cloth. There are two types of electric charge – positive and negative. Electrons have a negative electric charge. So when electrons move from cotton to plastic, the plastic gets a negative charge and the cloth gets a positive charge (because negative electrons leave it). The charges are not inside the object, but on the surface.

Attraction and repulsion

There is always a force between two electric charges. Like charges (two positive charges or two negative charges) repel (push each other away). Opposite charges (a positive charge and a negative charge) attract (pull towards each other). The force is quite small, but it can overcome **gravity**, lifting charged objects into the air. The two charges do not have to be touching to attract or repel each other. For example, when you comb your hair, opposite charges on the comb and your hair attract each other, making your hair stand on end.

Induced charges

An electric charge on one object can attract another object, even if the other object does not have a charge. For example, if a positively charged glass rod is held near a scrap of paper, electrons in the paper are attracted, creating a negative charge on the side of the paper nearest the rod, and a positive charge on the other side. The charges are called induced charges.

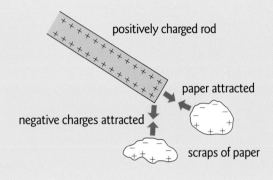

How a positive charge, such as that on a glass rod, attracts scraps of paper by induction.

positively charged rod

paper attracted

negative charges attracted

scraps of paper

Copying by static

Static electricity may be annoying when it makes your hair stick out, but it is also very useful, too. One of its many uses is in a photocopier.

Inside a photocopier there is a drum with a special outer coating. When you make a copy, the drum gets charged all over with a negative electric charge. The copier then shines light onto the original that you are copying. Where the original is lightly coloured, light bounces off, and on to the drum. Where it is dark, the light does not bounce off. Where light hits the drum, the special coating allows the charge to leak away. So a copy of the original is made in static electricity around the drum.

Where the drum is charged, it attracts fine black dust called toner. The toner is transferred on to fresh paper and heated up so that the toner melts to make the black areas of the copy on the paper.

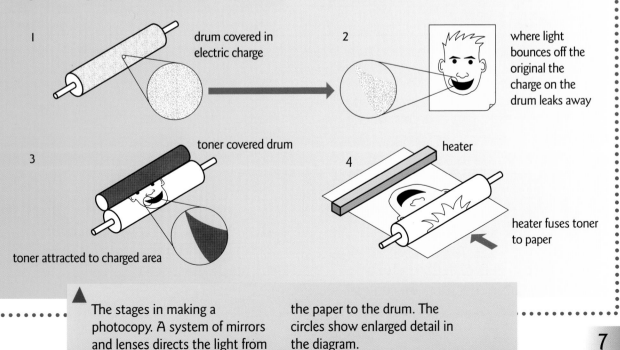

1 drum covered in electric charge

2 where light bounces off the original the charge on the drum leaks away

3 toner covered drum

toner attracted to charged area

4 heater

heater fuses toner to paper

The stages in making a photocopy. A system of mirrors and lenses directs the light from the paper to the drum. The circles show enlarged detail in the diagram.

Static on the move

If a charge is large enough and close enough to another object, it can jump through the air across the gap between the objects. As it passes through the air, it causes a spark of light and crackling noise, which you often hear when your clothes rub together. Lightning is a huge spark leaping between the ground and a thunder cloud.

SCIENCE ESSENTIALS

A moving electric charge is called an **electric current**.
A **conductor** is a material that electric charge can flow through easily.
An **insulator** is a material that electric charge cannot flow through.

Leaking away

If you touch a charged object, such as a ruler, on a large metal object, such as a radiator, the electric charge on the first object disappears. But, if you touch it against a wall, the charge remains.

This is because the charge can flow through the radiator, but not through the wall. The radiator is made of steel, which is called an electrical conductor because electric charge can flow through it. The wall is made of plaster, which is called an insulator because electricity cannot flow through it. The plastic ruler itself is also an insulator. Electric charge cannot flow through it, but can stay on the outside of it.

The Earth is like a huge 'sponge' that can soak up charge. Charge always tries to flow to Earth if it can. So when charge flows into a radiator, it is actually flowing into the ground along the radiator's pipes. When a charge flows to the Earth, the process is called grounding.

▲ In an engine spark plug, a huge charge leaps across the gap between the terminals, igniting the fuel in the engine's cylinder.

A flow of charge

When an electric charge moves through the air as a spark, or through a conductor, it is called an **electric current**. Water conducts electricity quite well (but not as well as metals), which is why dampness conducts away static electricity.

Lightning

The most dramatic example of jumping electric charges is the lightning that happens during storms. Inside a storm cloud there are millions of tiny ice crystals swirling up and down, hitting each other. This causes a huge electric charge to build up inside the cloud – a positive charge at the top and a negative charge at the bottom. Underneath the cloud, a positive charge builds up on the surface of the Earth as negative charges are repelled into the ground.

Sometimes a huge spark leaps from top to bottom of the cloud, causing 'sheet' lightning inside the cloud. Sometimes the charge at the bottom of the cloud leaps to the ground, creating 'fork' lightning. As the charge moves through the air, the air is heated up suddenly and expands, creating a thunderous boom.

▶ Lightning happens as positive and negative charges equal out inside a cloud, or between the cloud and the ground.

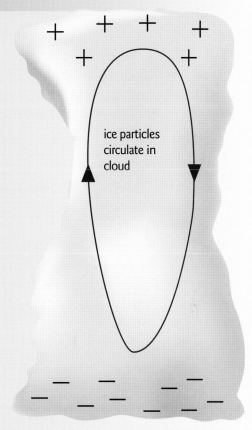

positive charge forms at top of cloud

ice particles circulate in cloud

negative charge forms at bottom of cloud

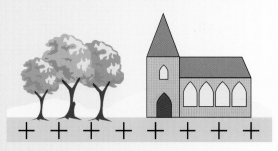

positive charge drawn up from ground

Lightning safety

Lightning always finds the easiest way to Earth. This means that objects which stick up – towers, trees and people – are in danger of being hit. If you are caught outside during a thunderstorm, keep low down and never stand under trees. Tall buildings have lightning conductors which are designed to be hit by lightning and carry the charge safely down to the ground.

Conductors and insulators

You know that an **electric current** is a flow of electric charge, and that electric current can flow through materials that are **conductors**, but not through materials that are **insulators**. Next, we will discover that some conducting materials are better than others.

Most metals, such as iron, copper, aluminium and silver, are very good conductors. Most other materials, such as plastics, **ceramics** and **organic** materials are insulators. Although insulators cannot conduct electric current, they are still useful in electric machines. For example, electrical cables are made by coating a wire, (which carries the current), with plastic. This protects the wire and stops the current escaping if the cable touches other wires.

Electrical safety

Electricity can be very dangerous. A large current flowing through a person can kill him or her. Your body is 90 per cent water, and water is a conductor, so electricity can flow through your body quite easily. This also means that you should never go near electrical machines with wet hands, and never take electrical machines into the bathroom.

Never go near overhead electrical power-lines or electricity sub-stations. The huge charges could jump through the air to you, even if you do not touch the machinery.

Conducting gases and liquids

Most conducting materials are solids, but some liquids and gases can conduct electricity, too – although they are not as good at conducting as solids. Air is a conductor because lightning and sparks can travel through it, but the electrical charges must be very high to create a spark. Water is a much better conductor because it contains charged particles which can move about in it.

SCIENCE ESSENTIALS

Most metals are good conductors.
Plastics are good insulators.
Some liquids and gases are also conductors.

► Cables at an electricity generating station, with insulating supports to stop the electricity jumping to the ground.

Free electrons

All materials are made up of **atoms**, and an atom is made up of a central **nucleus**, surrounded by **electrons**. Electrons have a negative electrical charge. Most **electric currents** are made up of a flow of electrons.

In some materials, there are electrons that are free to move from one atom to the next. These materials make good conductors because the electrons can flow easily through them. In other materials the electrons are firmly connected to their atoms, and cannot move. These materials are insulators because there are no free electrons to carry the current.

Semiconductors

A **semiconductor** is a material that can be either an insulator or a conductor. Semiconductors are used in many electronic devices, which you can find out about on page 16.

The most common material used to make semiconductors is the **element** silicon. It is extracted from silica, which is found in most types of rock – in fact, silicon makes up more than a quarter of the Earth's crust. On its own, silicon is an insulator. To make semiconductor devices, such as **diodes** and **transistors**, small amounts of other materials are added to the silicon. This is called doping.

There are two basic types of semiconductor – n-type and p-type. In an n-type semiconductor, the extra material (called an impurity) makes free electrons available. In a p-type semiconductor, the impurity removes electrons, creating holes that electrons can fit into. The types are combined to make semiconductor devices, such as the diode in the illustration, in which **resistance** can change from very high to very low automatically.

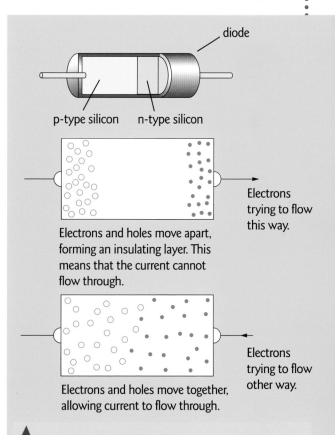

diode

p-type silicon   n-type silicon

Electrons trying to flow this way.

Electrons and holes move apart, forming an insulating layer. This means that the current cannot flow through.

Electrons trying to flow other way.

Electrons and holes move together, allowing current to flow through.

▲ A diode (top) is like a one-way street for electric current. It has a very high resistance (middle) to current flowing in one direction and a low resistance to current trying to flow in the other direction (bottom).

Electric current

An **electric current** normally flows around a loop of conducting material called an electric circuit. Electric current does not flow by itself. It needs a push to keep it going. This push is made by a difference in the charges of the two places between which the current flows, and is called an **electromotive force**. It can be made by a **battery** or by **mains** supply.

SCIENCE ESSENTIALS

An electric current normally flows in a loop of conducting material called a circuit.
A battery pushes the current around the circuit.

A simple circuit

There is a simple electric circuit inside a torch. It has three components – a battery, light-bulb and switch. They are joined together into a loop by strips of metal (a conducting material) inside the torch.

The battery pushes the current around the circuit. If there was no bulb, the current would just whiz around the loop. The bulb slows down the current as it contains a very thin wire which the current has to squeeze through.

The thin wire resists the flow of current, making it glow white hot.

The switch is used to break the circuit when you want to turn the bulb off. When the switch is closed, the circuit is said to be closed and the current flows around it. When the switch is open, the circuit is said to be open. The push from the battery is not large enough to make the current jump across the gap between the wires in the open switch, so the current cannot flow.

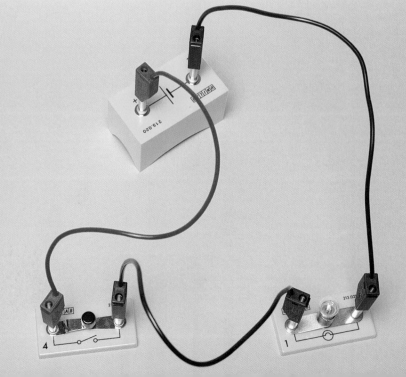

The components of a torch connected into a circuit by lengths of wire. They are a battery (top), a bulb (right) and a switch (left).

Cells and batteries

A cell is a store of energy. It contains chemicals that react together to make electric charges. These charges create an electromotive force which pushes current round a circuit. A battery is made up of two or more cells connected together in series or parallel. A battery makes more push than a single cell.

The chemicals in a battery are gradually used up as the battery gives out current. When they are used up, the battery is dead. Some types of battery are rechargeable and when current is put into them, the chemicals reform, ready for the battery to be used again.

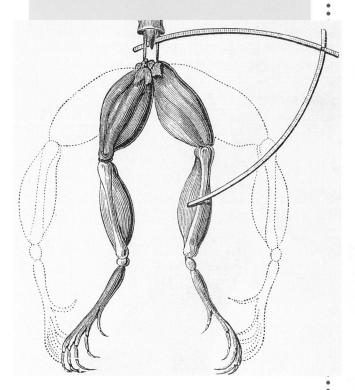

In this experiment, by Italian scientist Luigi Galvani in about 1780, a frog's leg twitched. Galvani's friend Alessandro Volta realised that this was caused by electricity made by two different metals being in contact, and went on to design the first cell.

Changing current

There are two types of electric current. Direct current (DC) is current that flows in just one direction around a circuit. Batteries make DC, so DC is used in simple circuits, such as those in a torch or a battery-powered fan.

Alternating current (AC) is current that keeps changing direction – it flows one way and then the other. Electricity generators make AC current, so AC is used in mains circuits.

Information in currents

A current that is able to change in strength and direction can carry information. For example, when you speak into a telephone, the pattern of sound that your voice makes is used to change the strength and direction of an electric current. This is done by a **microphone**. When the changing current is fed into a **loudspeaker**, the sound is reproduced. In this way, the sound can be sent along a wire as an **electrical signal**.

Current in circuits

Electric circuits contain components such as **batteries**, light-bulbs and **switches**, connected together with wires. The different combination of components, and the way that they are connected together makes the circuit do the job it is designed to do.

No matter how complicated a circuit is, **electric current** does not disappear anywhere. The same amount of current flows back into the battery or **mains** plug at the end of the circuit as leaves it at the beginning of the circuit. As the current flows around, it loses energy as it passes through the components.

SCIENCE ESSENTIALS

Electric current carries energy from place to place.
Current does not disappear as it flows around a circuit.
Current is measured in **amperes** (A).

Series and parallel

The components in a circuit can be arranged in series or parallel.
An example of a series circuit is in a lamp, where the electricity flows through the switch and then the bulb.
An example of a parallel circuit is the circuit which carries electricity to household sockets, where electricity is supplied to all the sockets at once.

▼ In the first circuit (left), current flows in series through the bulbs. In the second (right), it flows in parallel through the bulbs.

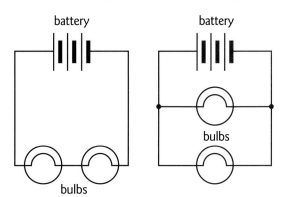

battery   battery

bulbs

bulbs

Measuring current and e.m.f.

The size of an electric current is the amount of charge that passes through a point in the circuit every second. It is measured in amperes (A) with an **ammeter**. The push that makes a current flow is called **electromotive force** (e.m.f.), and is measured in volts (V) using a voltmeter. The **power** of an electric current is measured in watts.

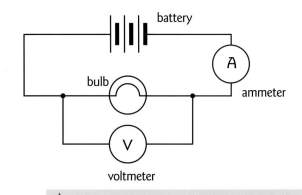

battery

bulb

A

ammeter

V

voltmeter

▲ An ammeter is connected in series with a circuit, and a voltmeter is connected in parallel.

Electrical resistance

Many electrical components reduce the flow of current in a circuit. They are said to have a **resistance**. The size of the resistance is measured in ohms (Ω). If you double the resistance in a circuit, keeping the electromotive force the same, the current is halved.

The smaller the resistance of a circuit, the higher the current that flows round it. If one part of a circuit becomes connected to another accidentally, reducing the resistance, a dangerously high current can flow and damage components in the circuit. This is called a short circuit. Fuses and circuit-breakers are automatic safety devices that cut off the electricity if the current gets too high.

Making light

One of the main uses of electricity is to make light, but how does current turn into light?

Incandescent bulbs
Incandescent bulbs are the type used in torches and as standard household bulbs. The current flows through a very thin wire of high **resistance**, making it so hot that it glows. The bulb contains an **inert** gas that stops the wire burning up.

Light-emitting diode (LED)
An LED is a special type of **diode** made with a **semiconductor** material that gives off light when current flows through it. LEDs need only a tiny current to make them work.

Arc lamps
In an arc lamp a very high voltage makes a continuous spark, called an arc, jump across a small gap in a circuit. This gives out a very bright light. Arc lamps are used for stage and film lighting or for strong floodlighting.

Light from gases
In a gas-discharge lamp there is a glass tube filled with gas. A high voltage makes the particles of gas divide into charged particles, which collide, making light. Different types of gas produce different colours of light.

Fluorescent tubes are similar, but the light is produced by the special **fluorescent** coating on the inside of the tube when it is hit by ultraviolet light created by the gas inside the tube.

▶ Gas-discharge tubes are often used for shop-window and advertising displays, as in this nighttime view of Hennesey Road, Hong Kong.

Electronics

Electronics are electrical circuits in which the components themselves control the flow of current. Machines with electronic circuits, such as hi-fis, videos, computers, electronic watches and radios, use electricity to keep time, do calculations, or **amplify** music automatically.

SCIENCE ESSENTIALS

Electronic circuits are electric circuits that control the flow of electricity.

Electric or electronic?

The diagram shows two circuits that do the same job, but one is electric and the other is electronic. With the electric circuit a person must close the switch to turn on the light-bulb when it gets dark. With the electronic circuit the light-bulb is turned on automatically.

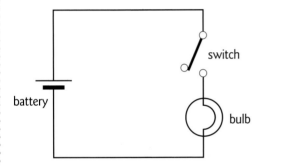

Electric circuit – close switch to turn on light-bulb.

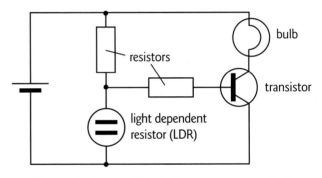

Electronic circuit – Light-bulb turns on automatically when light stops falling on LDR.

Electronic components

In electronic circuits the flow of electric current is controlled by electronic components. Each type of component does a specific job. A radio, a telephone and a computer will contain all these main types of components.

A **resistor** controls the size of the current that flows round a circuit or part of a circuit. Increasing the **resistance** reduces the current. Reducing the resistance increases the current. Most resistors have a fixed resistance, but the resistance of a variable resistor can be adjusted.

A **capacitor** stores electric charge on two metal plates that are separated by a sheet of insulating material. Current can flow in and out until the capacitor is full or empty, but then cannot flow through it. Some capacitors can store a huge charge, and some just a tiny charge.

A **diode** is a **semiconductor** that allows current to flow one way but not the other. It has a low resistance to current in one direction and a high resistance to current in the other direction.

A **transistor** is a semiconductor with three connections. The current that it allows to flow between two of the connections is controlled by the current that flows into the third connection. A transistor allows a tiny current to control a large current.

Optical devices

Some components are controlled by the amount of light that falls on them. For example, a light-dependent resistor has a very low resistance in bright light and a very high resistance in the dark.

*T*ransistor times

The development of the transistor was the most important event in the history of electronics. The first transistor was made in 1948 at Bell Laboratories in the USA.

Before semiconductors were developed, there were electronic devices called valves to do similar jobs, but they were contained in glass bulbs and were several centimetres long – hundreds of times bigger than modern semiconducting devices. They also used far more electrical energy and regularly broke down.

The transistor gradually replaced a type of valve called a triode, allowing electronic devices such as radio, television and computers to be much smaller, cheaper and more reliable.

▼

Early computers, such as the ENIAC (Electronic Integrator and Computer), built with valves rather than microchips, were enormous, yet less powerful than today's pocket calculators. ENIAC, completed in 1946, had about 18,000 valves.

Microchips

Some electronic circuits are built up of different components connected by wires. Each component, such as a **resistor** or **transistor**, is in its own casing and has its own connections. These components are called discreet components.

A microchip, properly called an **integrated circuit**, is a tiny wafer of **semiconductor** material that contains a complete electronic circuit, with the components built into it.

The components are so small they can only be seen through a microscope.

Most microchips come in the form of a black plastic rectangle with metal legs sticking out. The plastic is a case that protects the delicate integrated circuit inside and the legs carry electricity to and from the integrated circuit. On the integrated circuit each miniaturized component is just a thousandth of a millimetre across.

Chip types

Microchips can be designed to do different jobs. Some chips have quite simple circuits with just a few connections. For example, there are chips that contain amplifying circuits, chips that work digital watches, and even chips that contain complete radios. Other chips have millions of components and do very complex jobs. Examples are computer **memory** chips and **microprocessor** chips.

History of the chip

The first integrated circuit was made by the company Texas Instruments in the USA in 1959. It contained just a few components. Since then it has become possible to fit more and more components on a chip, and chips have become very cheap.

One example of the increasing power of chips is the microprocessor, which is at the heart of every personal computer. The first microprocessor chip was the Intel 4004, built in 1972. The first 'home computers', which appeared around 1980, had microprocessors that did calculations on 8-bit numbers, working at a speed of 8 megahertz (8 million cycles per second). Currently, PC processors are working with 32-bit numbers at more than 500 megahertz. This is an increase in computing power of several hundred times.

In 1980, a typical personal computer had 8 kilobytes of **RAM**. A modern PC can have hundreds of megabytes of RAM. That's tens of thousands of times more memory in the same space and for the same price!

Making chips

Making microchips, with their microscopically small components, is a very complex job that must be carried out in purpose built, dust-free factories. Chips start life as a thin, circular wafer about 15 centimetres across, cut from a cylinder of pure silicon. The components and connections between them are built up using photographic techniques and chemical treatments. The wafer is then broken up into individual chips, which are encased in plastic and have their connecting legs added.

▼ Electronic engineers inspecting a plan of a microchip. The final chip will be less than the size of one of their fingernails

Digital circuits

With electronic components such as **transistors**, **diodes** and **capacitors**, it is possible to build circuits called digital circuits. In these circuits the **binary numbers** 1 and 0 are represented by a high voltage or a low voltage. There are circuits that can remember a 0 or a 1, (called memory circuits), circuits that count, and circuits called logic gates. By combining these circuits together, larger circuits that can do complex digital calculations can be built.

> ▼ Logic gates are electronic circuits which make simple decisions. The output depends on the combination of the inputs. A truth table for a logic gate shows what the output is for different inputs

Logic gates

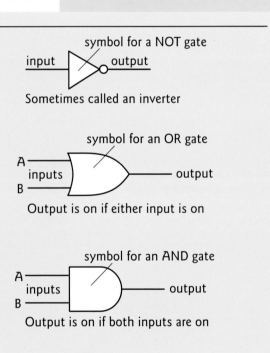

Logic gate	Truth table		
NOT	Input	Output	
	0	I	
	I	0	

symbol for a NOT gate

input — output

Sometimes called an inverter

OR	Inputs		Output
	A	B	
	0	0	0
	0	I	I
	I	0	I
	I	I	I

symbol for an OR gate

A — inputs — output
B

Output is on if either input is on

AND	Inputs		Output
	A	B	
	0	0	0
	0	I	0
	I	0	0
	I	I	I

symbol for an AND gate

A — inputs — output
B

Output is on if both inputs are on

A logic gate is a circuit that makes simple decisions. It has inputs and outputs. The **voltage** (which represents 0 or 1) at the output depends on voltages (which represent 0s and 1s) at the inputs. By combining these simple circuits, it is possible to make circuits that can do simple sums with binary numbers, such as adding and subtracting.

The simplest logic gate is the NOT gate, which has one input and one output. It turns 0 into 1 and 1 into 0. An OR gate has two or more inputs and one output. If there is a 1 at any of the inputs, there is a 1 at the output. An AND gate also has two or more inputs and one output. If there is a 1 at all the inputs, there is a 1 at the output.

You might use an OR gate in a burglar alarm system. A signal from each sensor would go to the inputs of the gate and the output would trigger the alarm bell. If any one of the sensors was triggered by a burglar, the OR gate would set off the alarm.

You might use an AND gate in the safety system of a lift, with an input from both sets of lift doors (the inside and outside sets). Only if both sets were shut would the output tell the lift to start.

Digital communications

You have probably heard about the digital broadcasting of television and radio, and digital telephone systems, but why are broadcasting and telecommunications companies introducing these systems? The answer is that they make the communications much clearer and more efficient.

Original communications systems were called analogue. This means that, for example, sound is represented by an electric current constantly changing strength and direction.

In digital systems the analogue signal is digitized, which means it is turned into a list of binary numbers. Electronic circuits keep measuring the strength of the current and turn the measurements into binary numbers. Because numbers only have 0s and 1s in them, they can be sent as simple signals by turning the current on or off. These are called digital signals.

When an analogue signal gets slightly distorted, the quality of the sound or picture it represents is reduced.

However, even if a digital signal is distorted, the ons and offs are still readable, so the quality is not affected. As digital signals are more simple than analogue signals, more of them can be sent along a cable at once, making digital communications more efficient.

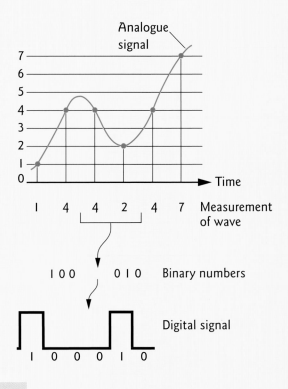

How part of an analogue signal (top) is converted into a simple digital signal (bottom). In a broadcasting system, measurements are taken thousands of times per second.

Magnets

Magnets are objects which attract (pull on) iron and steel objects without touching them. We use magnets around our homes for jobs such as keeping doors shut, sticking notes to a refrigerator door and for keeping paperclips together. There are many more magnets hidden away inside gadgets and machines.

SCIENCE ESSENTIALS

Materials which are attracted by magnets are called **magnetic** materials.
The region around a magnet is called a magnetic field.
The direction of the field is shown by lines of force.
Like poles repel and opposite poles attract.

Magnetic materials

Magnets only attract objects that contain certain materials, called magnetic materials. The most common magnetic material is the metal iron. **Steel** is also magnetic because it is composed mostly of iron. Other metals, such as copper and aluminium, and materials such as plastic, **ceramic** and wood are not magnetic. Magnets themselves are made of magnetic materials. The best magnets are made from steel **alloys** – materials made by mixing small amounts of other materials, such as aluminium and nickel, with steel.

Magnetic fields

The area around a magnet where its magnetism can be felt is called a **magnetic field**. It stretches all around the magnet. It is strongest near the ends of a magnet and quickly gets weaker further from the magnet. This means that the closer a magnetic object is to the magnet, the stronger the magnet pulls on it.

▶ Although a magnetic field is invisible, it can be indicated with lines of force showing its direction.

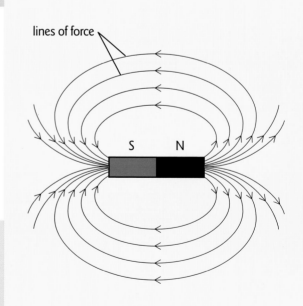

lines of force

Magnetic poles

The magnetic field around a magnet is always concentrated in two places on the magnet, no matter what size or shape the magnet is. These are called **magnetic poles**. One is always called the north pole (and is labelled N) and the other is always called the south pole (and is labelled S).

A magnet always attracts a piece of magnetic material which is not a magnet. However, when two magnets are in each other's fields, their poles interact. Like poles (two south poles or two north poles) repel each other; opposite poles (a north pole and a south pole) attract each other.

What causes magnetism?

Scientists think that magnetism is caused by the movement of **electrons** – the tiny, negatively charged particles in atoms which move around the atom's **nucleus**. In most materials, the effects of the movement cancel each other out, but in magnetic materials they combine to form tiny magnetic regions called domains, which act like tiny **bar magnets**.

In magnetic material which is not magnetized, the domains point in random directions and their magnetism cancels each other out. In a magnet, all the domains are lined up with their north poles pointing in the same direction.

▼ A piece of magnetic material, such as a steel paper clip, becomes a magnet when it is near a permanent magnet. This is called induced magnetism.

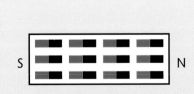

Unmagnetized material

Domains cancel out

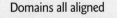

Domains all aligned

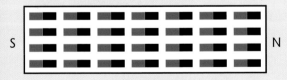

S   N   S   N

Permanent magnet

When brought near a magnet, the domains turn so that the material itself becomes a magnet.

Earth's magnetism

You probably know that you can use a compass to find your way in the countryside. But how does a compass work? The answer is that is uses the Earth's **magnetic field**. This field is created because the Earth acts like a huge **bar magnet**, with its poles near, but not in the same place as, the Earth's **geographic poles**.

SCIENCE ESSENTIALS

The Earth has a magnetic field as though it had a bar magnet inside. A compass needle aligns itself between magnetic north and magnetic south, so that one end of the needle points north and the other points south.

Earth's magnetic field

The Earth's magnetic field is the same shape as the field around a bar magnet. The lines of force come from one pole, come out of the Earth's surface, curve round through space and then re-enter the Earth to reach the other pole. The Earth's field stretches almost 100,000 kilometres out into space.

▶ Many species of birds can sense the Earth's magnetic field and use it to find their way as they migrate.

Compasses

A compass contains a needle that is a tiny bar magnet which is free to swing round. When the compass is held level, the needle swings to point to the Earth's magnetic poles. It indicates magnetic north and magnetic south. Because a compass always points in the same direction, it can be used to follow a straight line on the Earth's surface.

Compasses can sometimes be confusing. Near the equator magnetic north is almost in the same direction as geographic or 'true' north, but if you looked at a compass at the geographic North Pole, it would try to point to the south and would not be much good at all. Some rocks, such as gabbro, contain lots of iron and are magnetic. If you try to read a compass where these magnetic rocks are underground, the rocks will attract the compass needle and make the compass reading unreliable.

North and south poles

All magnets have two poles called north and south. These names are shortened versions of 'north-seeking pole' and 'south-seeking pole'. They have these names because if you suspend a magnet on a thread so that it can spin, one pole always turns to face magnetic north (the north-seeking pole) and the other turns to face magnetic south (the south-seeking pole).

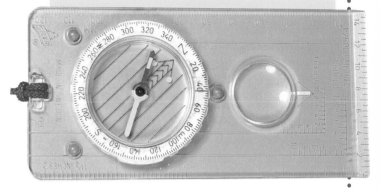

A walker's compass. The red end of the needle is the north-seeking pole, which always points to the north magnetic pole.

Currents of iron

We know that an **electric current** or moving **electrons** cause magnetic fields, but what makes the Earth act as a magnet? Without travelling deep inside the Earth, it is difficult to find out. Geologists think that the cause is linked to the fact that the Earth's core is made of iron. However, the iron is extremely hot and molten, which means that it cannot be a solid magnet. Another theory is that there are currents of iron swirling around in the liquid core, making it act like a giant **electromagnet**.

Moving poles

The magnetic poles are not only in a different place to the geographic poles, but they are also slowly moving. The magnetic axis (the line between the two magnetic poles) wobbles around the geographic poles. More dramatically, the two magnetic poles have swapped places many times over the last few million years. We know this because the direction of the magnetic field is recorded in ancient **igneous rocks**. As the solid rocks were formed by molten rocks cooling, the magnetic parts of the rocks, such as iron, were lined up with the magnetic field.

The movement of the poles is important when navigating by map and compass. A map always shows the direction to magnetic north, but after a few years this can be quite inaccurate. To cope with this problem, the change of angle each year is always written on the map.

Electric magnets

Magnetism is closely related to electricity. In fact, whenever a current flows, a **magnetic field** is created around the wire that the electric current is flowing through. This effect is called electromagnetism and a magnet made in this way is called an **electromagnet**.

SCIENCE ESSENTIALS

An electric current makes a magnetic field.
An electromagnet is a magnet made by electricity flowing through a coil of wire.
An electromagnet has a magnetic field, just as a bar magnet has a magnetic field.

Electromagnets

Although an electric current in a wire creates a **magnetic field** around the wire, the field is extremely weak and much weaker than the Earth's magnetic field. If the wire is wound round and round to make a coil, the magnetic field is much stronger. By putting an iron core in the middle of the coil it can be made stronger still. A coil of wire with an iron core acts like a **bar magnet**, with a pole at either end.

The strength of the magnetic field made by a an electromagnet depends on the number of turns of the wire in the coil (more turns makes a stronger magnet) and the current flowing through the wire (bigger current makes a stronger magnet). The position of the north and south poles depends on which way round the wire is coiled and which way the current flows.

A simple electromagnet, made by winding a long length of wire around an iron rod. It becomes a magnet when a current flows through the wire.

Advantages of electromagnets

Although electromagnets need an electric current to make them work, while permanent magnets do not, they have several advantages over permanent magnets. Their magnetism can be turned on and off just by turning the current on and off. The strength of their field can be adjusted by adjusting the strength of the current, and they can be made much stronger than permanent magnets. The north and south poles can also be swapped over just by reversing the current. This makes electromagnets useful for turning electrical energy into movement energy.

Transforming electricity

When the current flowing through a wire changes strength or direction, the magnetic field around the wire changes. The reverse of this happens, too. If a wire is in a magnetic field that changes strength or has poles which swap round, an electric current is set up in the wire. This effect is called **electromagnetic induction**, because current is induced in the wire.

An electrical device called a transformer makes use of electromagnetic induction. The job of a transformer is to transfer current from one wire to another without the wire actually being connected. A simple transformer has an iron core with two coils of wire around it. When the current in one coil changes, it creates a changing magnetic field in the core, which in turn creates an electric current in the other coil.

Transformers are normally used to increase or decrease the **voltage** in a wire. For example, the voltage of **mains** electricity is 100 volts or 220 volts.

This is too high for the electronic circuits in many machines, such as computers and radios, so a transformer is used to reduce the voltage to, say 9 volts.

▼ An industrial transformer. You can see the huge coils of wire and iron core linking them together.

Movement with magnets

One of the most important applications of **electromagnets** is to create movement from electrical energy. Turning on an electromagnet makes it attract an object made of **magnetic** material, or **attract** or **repel** a **permanent magnet** or another electromagnet. Of course, a permanent magnet can also do this, but it cannot be turned off again to stop attracting or repelling the object.

SCIENCE ESSENTIALS

An electromagnet can turn electrical energy into movement energy.
An electric motor uses electromagnets to create turning movement.
A dynamo or alternator creates electricity when it is spun round.

Simple movement

A simple machine such as an automatic door latch uses an electromagnet to move one of its parts, in this case to pull back the latch so that the door can be opened. In an electric motor, electromagnets create a continuous spinning motion of the motor's spindle. Attached to the spindle is an electromagnet called an armature, surrounded by permanent magnets. When electricity is sent around the armature, first one way and then the other, the armature is continuously attracted to the magnets, making it spin.

Electricity from movement

A dynamo is very similar to an electric motor, but it changes movement into electricity. Spinning its spindle makes it create an electric current by electromagnetic induction. For example, a bicycle's dynamo creates electricity for the cycle's lights as the wheel turns.

A large electric motor such as the one being repaired here would be used to power a large machine, such as an electric train.

Electricity and magnetism in our lives

You have seen that electricity is a form of energy that is easy to distribute to homes, offices and factories, and that when it arrives, it is easy to convert into other forms of energy, such as heat and light. The fact that electromagnets can turn electrical energy into movement makes electricity even more useful. Electronic components make it possible to build a range of complex electronic machines, from radios to computers, as well as circuits that control other machines, from washing machines to aircraft.

Electromagnets in communications

Electromagnets are an important part of communication and hi-fi equipment, such as **microphones**, **loudspeakers** and tape recorders. In all these pieces of equipment, sounds are represented by a changing **electric current** called an **electrical signal**.

In a type of microphone called a moving-coil microphone, sound hits a **diaphragm** attached to a coil of wire, which has a permanent magnet inside it. The sound vibrates the coil, which sets up an electrical signal in the wire.

In a loudspeaker, an electrical signal is fed to a coil of wire with a permanent magnet around it, attached to the centre of a cardboard cone. The changing electric current makes the coil vibrate, which makes the cone vibrate, producing sound.

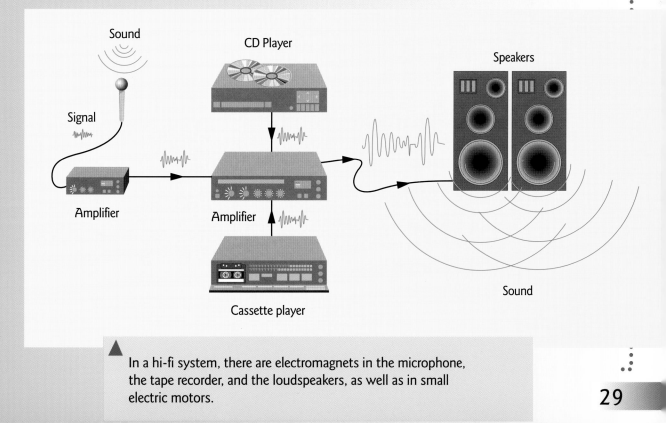

In a hi-fi system, there are electromagnets in the microphone, the tape recorder, and the loudspeakers, as well as in small electric motors.

Glossary

alloy material made by mixing a metal with another substance, which can also be a metal. Steel is an alloy of iron and carbon.

amber yellow solid which is the fossilized sap of prehistoric trees

ammeter a device which measures the strength of an electic current in amperes.

ampere (A) a unit of electric current

amplifier electrical device which increases the strength of an electrical signal. A **transistor** can act as an amplifier.

atom one of the tiny building blocks that all substances are made from. An atom is made up of a central nucleus surrounded by **electrons**.

attraction force that tries to pull two objects together

bar magnet magnet in the shape of a cylinder or rectangular block, with a pole at each end

battery store of electrical energy in the form of chemicals, which can push an electric current around a circuit

binary numbers numbers that are shown using a combination of the digits 0 and 1

bit single binary digit, either 0 or 1

byte part of the computer memory that stores a single character of data, usually eight **bits**. For example, the letter 'F' can be stored as 01000110 (as a **binary number**).

capacitor electronic component that stores electric charge

ceramics family of materials made by baking clay

conductor material that allows **electric current** to flow through it easily

current electricity electric charge that flows

diaphragm sheet of material that can vibrate up and down, like the skin of a drum

diode **semiconductor** electronic component that allows electric current to flow one way, but not the other way

electric current flow of electric charge

electrical signal **electric current** that continuously changes in strength and direction to represent information

electromagnet magnet made by an **electric current** flowing in a coil of wire

electromotive force (e.m.f.) push that is needed to make an electric current flow. It is normally made by a battery or mains supply.

electron one of the tiny particles that make up an **atom**

element elements are the simplest substances that exist. An element is made up of just one type of **atom**.

fluorescent describes a material that gives off light when it is hit by certain types of radiation, such as ultraviolet light

geographic poles two points on the Earth's surface which are on the axis that the Earth spins round

gravity force that attracts every object to every other object. Gravity pulls you towards the Earth.

igneous rock rock that forms when hot, molten rock cools and turns into solid rock

inert describes a substance that does not react chemically with any other substance. Gases such as neon and argon are inert gases.

insulator material that does not allow **electric current** to flow through it

integrated circuit electronic circuit consisting of microscopically small components built onto a tiny chip of semiconducting material such as silicon

loudspeaker device that turns an electrical signal into sound using an **electromagnet**

magnetic describes a material that is attracted by a magnet. The most common magnetic material is iron.

magnetic field region around a **magnet** where its effect can be felt

magnetic poles two places on a magnet where the magnetic field is most concentrated, called the north pole and south pole

mains supply of electricity to homes, offices and factories from electricity generating stations

memory part of a computer where information is stored electronically in memory chips

microphone device that turns the patterns of a sound into an **electrical signal**

microprocessor integrated circuit that can perform calculations and other functions under instruction by a program. There is a microprocessor at the heart of every computer.

nucleus central part of an atom

organic describes substances that are parts of plants or animals, or that are made from plants or animals

permanent magnet piece of magnetic material that is a magnet all the time. Other magnets are temporary magnets.

power amount of energy given out or used up every second

RAM short for Random Access Memory, type of computer memory

repulsion a force which tries to push two objects apart

resistor electronic component that resists the flow of electric current

resistance amount by which a **resistor** or other component tries to resist the flow of an **electric current**

semiconductor substance that can act as both a **conductor** and an **insulator**

static electricity electric charges that cling to the outside of insulating objects; made by rubbing objects together

switch device that completes or breaks an electric circuit

transistor **semiconductor** electronic component that can act as an **amplifier** or an electronic switch

voltage push that makes an **electric current** flow around a circuit or through a component, measured in volts (V)

KNOWSLEY HEY SCHOOL LIBRARY

Index